Published 1982 by Warwick Press,
730 Fifth Avenue, New York, New York 10019.

First published in Great Britain by
Kingfisher Books Limited, 1981.

Copyright © 1981 by Kingfisher Books Limited

Printed in Italy by Vallardi Industrie Grafiche.

6 5 4 3 2 1 All rights reserved.

Library of Congress Catalog Card No.81-51796

ISBN 0-531-09187-2

# First Picture Book of Animals

Written by David Lambert

Designed by Dave Nash

Illustrated by Mike Atkinson, John Francis,
David Hurrel, Bernard Robinson, Ralph Stobart and
David Wright

**WARWICK PRESS**

# Contents

| | |
|---|---|
| Moving About | 8 |
| Staying Alive | 10 |
| Animal Senses | 12 |
| Living Together | 14 |
| Insect Communities | 16 |
| Finding a Mate | 18 |
| Animal Homes | 20 |
| Master Builders | 22 |
| Looking after their Young | 24 |
| The Winter Sleep | 26 |
| Animal Journeys | 28 |
| Small Creatures | 30 |
| River Life | 32 |
| Snakes and Lizards | 34 |

| | |
|---|---|
| Animals of the Dark | 36 |
| Animals of Shallow Seas | 38 |
| Ocean Animals | 40 |
| All Kinds of Birds | 42 |
| Elephants | 44 |
| The Big Cats | 46 |
| Bears, Pandas and Koalas | 48 |
| Apes and Monkeys | 50 |
| Animal Oddities | 52 |
| Animals of Long Ago | 54 |
| The Animal Kingdom | 56 |
| Quiz | 58 |
| Index | 60 |

# Moving About

Some animals move only once in their lives. Baby oysters and sponges, for example, swim until they find a good place to stay. Then they fasten themselves on to rocks and never move again. But most creatures travel to find food, warmth or a mate. They move to escape enemies, heat, frost or floods.

To move in any direction, an animal must push something in the opposite direction. A horse trots by pushing back against the ground. A bird flies forward by pushing air backwards. A fish swims forward by thrusting water back.

Animals have bodies which help them move whether they walk, swim or fly. Like cars, planes or ships, animals have "engines" giving them the power to move themselves from place to place. A creature's engines are its muscles. Like a car's engine, muscles burn up fuel. An animal's fuel is the food it eats.

Pistons move a car by turning wheels, but muscles move most animals by pulling levers. These levers are the limbs. Muscles tend to work in pairs. As one muscle shortens, the other in the same pair lengthens, and so on. Pairs of muscles fixed to long hinged bones move them to and fro. Limb bones worked like this help a dog to run or a bird to flap its wings. Dogs push back with each limb, then bring each limb forward ready for another push. When a bird flaps its wings or a kangaroo leaps, the limbs work together as a pair. But when a horse or a dog walks or runs, the limbs work in turn.

Not all animals have legs or wings to help them travel. A squid shoots itself along by squirting water out of its body. A fish swims along by bending its body from side to side and thrusting water back with its tail. A clam has one foot that it uses to dig into mud or sand. A slug slides forward by pushing against the ground with its muscular foot. Worms have no legs at all. They use bristles and plates on their skin to pull themselves along the ground.

▼ **The pictures show how a horse moves from walking to trotting to galloping. In a gallop only two feet touch the ground at once.**

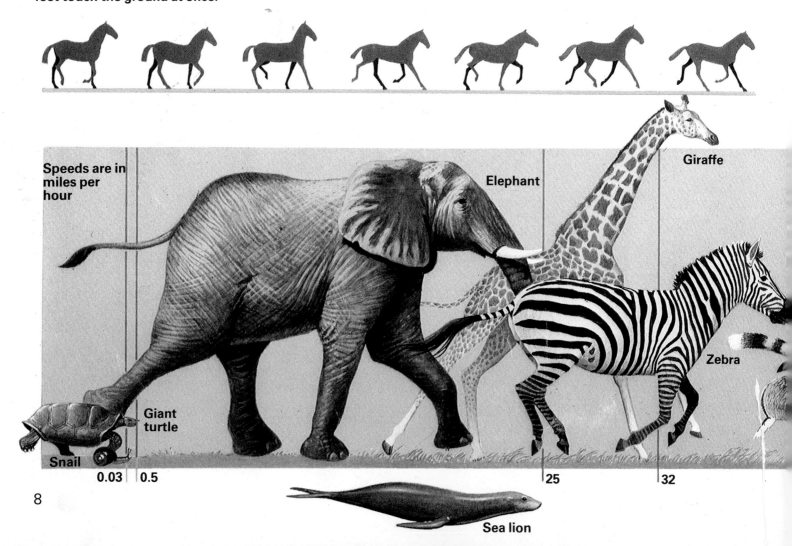

Speeds are in miles per hour

Snail 0.03 | Giant turtle 0.5 | Elephant | Giraffe | Zebra 32 | 25

Sea lion

▶ A bird flaps its wings as it flies. As its wings thrust back, the feathers push air back and the bird forward. Then the bird brings its wings forward again. This time the feathers let the air through.

▲ A looper caterpillar throws its body into a loop to bring its back legs forward. Then its front moves on.

▼ A fish bends its body to swim forward.

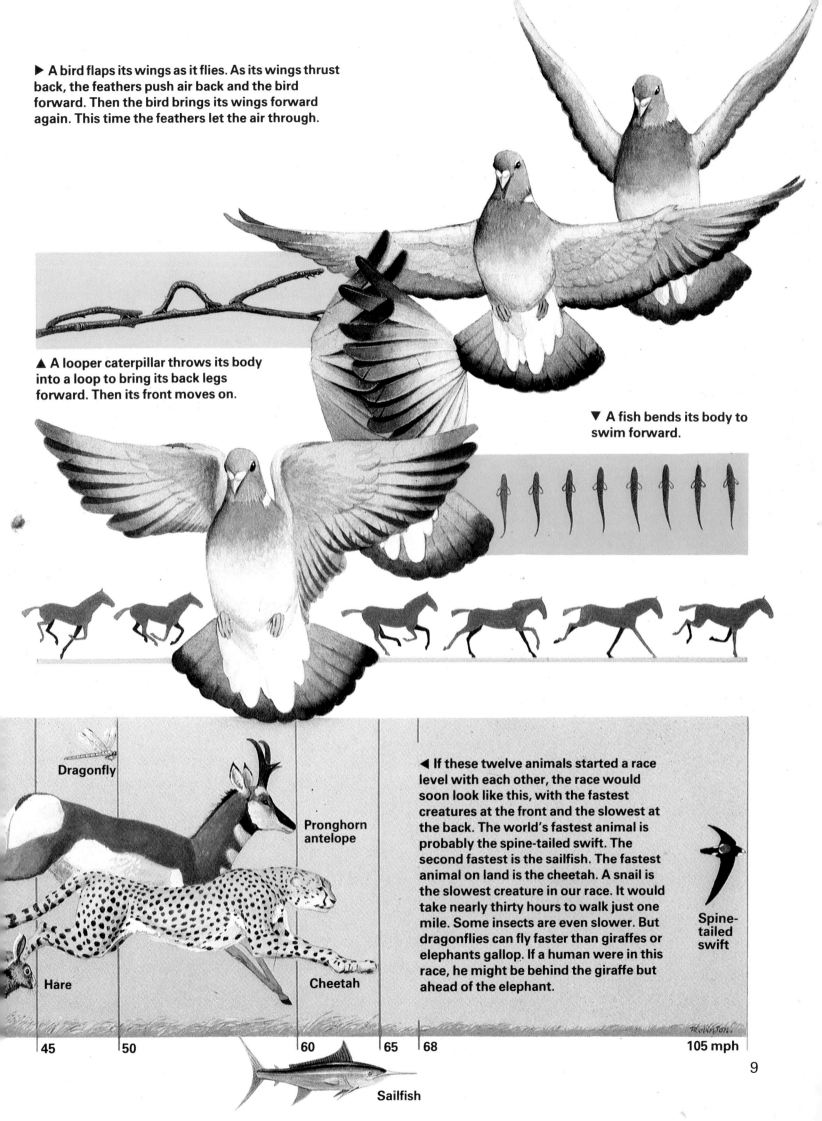

Dragonfly

Pronghorn antelope

Hare

Cheetah

Spine-tailed swift

◀ If these twelve animals started a race level with each other, the race would soon look like this, with the fastest creatures at the front and the slowest at the back. The world's fastest animal is probably the spine-tailed swift. The second fastest is the sailfish. The fastest animal on land is the cheetah. A snail is the slowest creature in our race. It would take nearly thirty hours to walk just one mile. Some insects are even slower. But dragonflies can fly faster than giraffes or elephants gallop. If a human were in this race, he might be behind the giraffe but ahead of the elephant.

45  50  60  65  68  105 mph

Sailfish

# Staying Alive

All animals must eat to live. Many eat plants but some eat other animals. These are the hunters.

Hunters all have special weapons or powers which help them catch the animals they eat. Cats have sharp claws and teeth. Eagles have curved claws to seize their prey and strong, hooked beaks for tearing off chunks of flesh. Some hunters, like wolves and cheetahs, can run fast to catch their prey. Others, like the ant lion, make traps to provide themselves with food.

Hunted animals all need some kind of protection to keep them from being eaten. Some are fast runners. If danger threatens, deer and antelope can run away faster than most enemies can chase them. Other animals, like turtles and porcupines, move too slowly to escape. But they have armor to protect them. A turtle can hide in its shell and a porcupine bristles with spines.

Some animals are good fighters. They attack the creatures that threaten them. A rhinoceros butts with its horns and a baboon will rush an enemy and strike with its teeth.

Then there are animals colored in ways that hide them from their enemies. A green caterpillar's color helps to hide it from birds while it munches green leaves. But color helps hunters too. In winter the Arctic fox turns white to match the snow. When the fox keeps still, the birds and mammals that it hunts cannot see it.

Hunted animals are often killed only when they become too old or sick to hide, fight or run away.

▲ A porcupine's sharp spines are loose enough to come off and stick into enemies that attack it.

▼ Ant lions are insects that dig pit traps in sand. They kill small creatures which fall in.

▼ The armadillo has bony plates on its back, head and tail. It protects itself by curling into a ball.

▲ Horns and a heavy, thick-skinned body protect the rhinoceros from enemies.

▲ A golden eagle swoops down, ready to sink its hooked claws into its victim.

▲ The small fierce piranha lives in South American rivers. It has sharp pointed teeth like little knives.

◄ Arctic foxes change color with the seasons to blend with the color of the land.

Summer coat

Winter coat

► Its speed may help a cheetah to catch an impala. But a herd of impalas also moves very fast. Their white and brown colors flash as they leap away and confuse the cheetah.

# Animal Senses

Just like us, animals have eyes, ears and other senses to tell them what is happening in the world around them. They have senses to help them find food and companions and to avoid danger.

Different animals have different kinds of eyes. Flatworms have simple eyes that only see light and dark. The eyes of crabs and flies are made up of many tiny eyes. These eyes are good at spotting any nearby movement. But they form a fuzzy picture of what they see. Birds and mammals have eyes with *lenses* that can change shape or *focus*. Eyes like this can form clear pictures of objects whether they are nearby or far away. Some birds have very sharp eyesight to help them hunt.

Ears hear sounds set up by vibrations in air or water. Some animals' ears seem strangely placed to us. A cricket's "ears" for example, are on its knees! Many animals can pick up sounds too high for human ears to hear. Bats and porpoises make high-pitched squeaks. If something is in front of them, the squeaks make an echo and the animals fly or swim around it.

Worms feel vibrations in the ground. Fishes sense vibrations with special hairs which grow in rows of tiny tubes along the sides of their bodies. This is called the *lateral line system*.

Smell and taste are important senses too. Dogs can pick up smells too faint for our noses to sense. Squirrels can smell whether nuts are good or rotten before they crack them open. Insects smell with feelers called *antennae*. Some insects can taste sugar with special hairs on their feet.

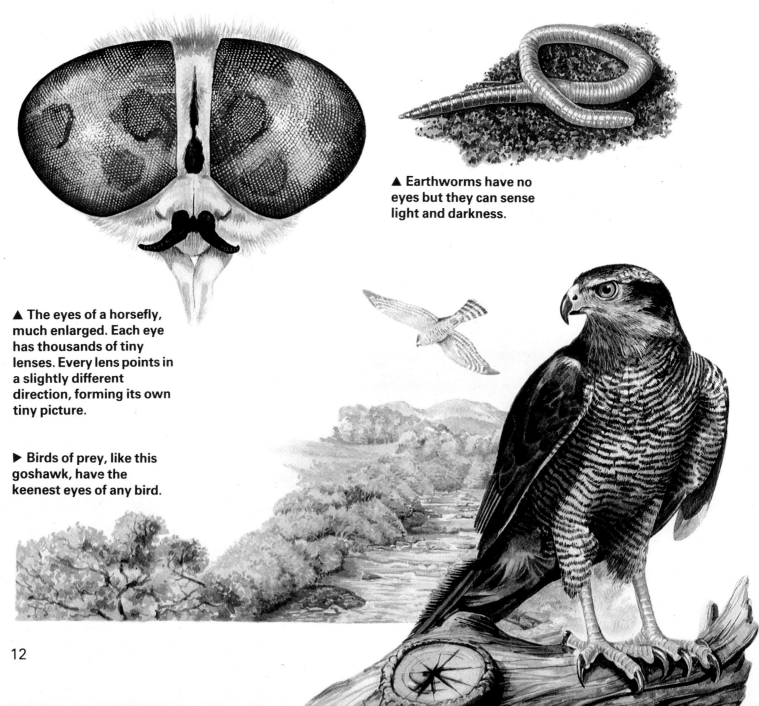

▲ The eyes of a horsefly, much enlarged. Each eye has thousands of tiny lenses. Every lens points in a slightly different direction, forming its own tiny picture.

▲ Earthworms have no eyes but they can sense light and darkness.

▶ Birds of prey, like this goshawk, have the keenest eyes of any bird.

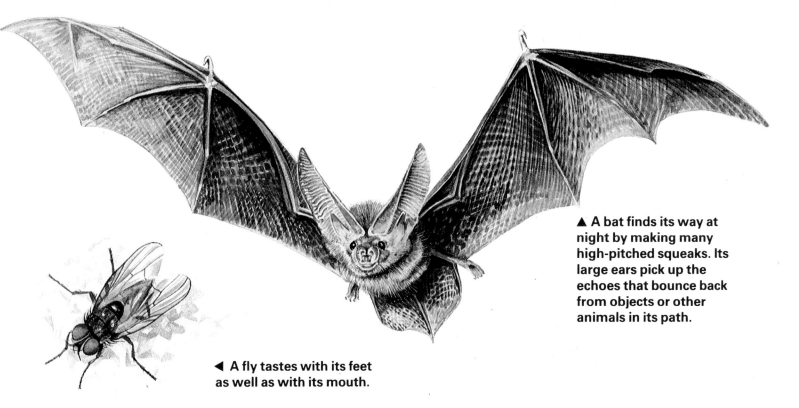

▲ A bat finds its way at night by making many high-pitched squeaks. Its large ears pick up the echoes that bounce back from objects or other animals in its path.

◀ A fly tastes with its feet as well as with its mouth.

Nerves that end in our skin can sense heat, cold, pressure, pain and touch. Insects have hairs that work rather like this. Some snakes called pit vipers sense heat with special pits just below their eyes. The pits pick up the warmth of birds and mammals. So pit vipers can hunt at night when it is too dark for them to see.

Sense organs like eyes, ears and noses work in a similar way. They all depend on nerves with endings on the outside of a creature's body. If a light, a sound or a scent reaches these endings, the nerves send electric signals to the creature's brain. The animal's brain then decides what the signals mean and how to react to them.

▼ A snake's tongue helps it smell things.

▼ The hairs in the line of special cells along the side of a fish warn it when other animals are near.

▼ A male Moon moth can smell with its feelers a female moth more than a mile away.

▶ The fennec fox is the smallest fox but it has the largest ears. This means that it can hear the slightest sound in its desert home.

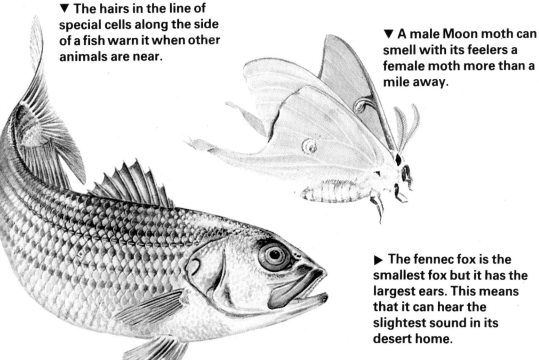

# Living Together

Many animals live alone. But some share their lives with others. This may help them to find food or avoid danger. Antelopes and other grazing animals, for example, roam the plains of Africa in herds. While most of them feed, a few watch for enemies like lions. Some hunters also live together. Wolves and wild hunting dogs form groups called packs. By attacking in packs they can kill deer or other animals much larger than themselves.

Some fishes, such as mackerel and herrings, swim in huge *shoals* for protection. A twisting shoal of mackerel can make a difficult target for the porpoises and other animals that hunt them.

Those graceful seabirds, Arctic terns, nest in groups called *colonies*. If a fox or a stoat approaches the colony, the birds swoop together to the attack. A flock of terns may drive off an enemy too big to be scared away by a single bird.

Sometimes animals of different kinds will live together. In the coral seas around Australia, for example, the clownfish lives among the tentacles of a sea anemone. Sea anemones have tentacles that fire off stinging darts and kill small fishes brushing past them. Yet the clownfish can hide among these tentacles unharmed. The sea anemone protects the clownfish from its enemies. In return, the clownfish drops scraps of food which the sea anemone eats. In this way each creature helps the other.

Other unlikely partners are the Nile crocodile and the spur-winged plover. A crocodile will lie with open jaws and let the bird walk in its mouth. The plover probably eats leeches fastened to the crocodile's tongue. Leeches are worms that suck blood. By eating the leeches, the plovers get a meal and in return they rid the crocodile of unwelcome *parasites*. Parasites are animals or plants that live and feed on others. In the same way, birds called oxpeckers feed on ticks that burrow in the hides of the buffalo and rhino.

Zebra  Ostrich  Wildebeeste

▼ Clownfishes in the tentacles of a sea anemone.

◀ Nesting terns attack a stoat. The birds scream as they swoop at the intruder's head. Terns normally just scare their enemies away, but their beaks can stab a young stoat to death.

▼ On the African plains, the big grazing animals often band together into herds for protection. In the front of this picture you can see a spur-winged plover searching for a meal on a Nile crocodile's back.

# Insect Communities

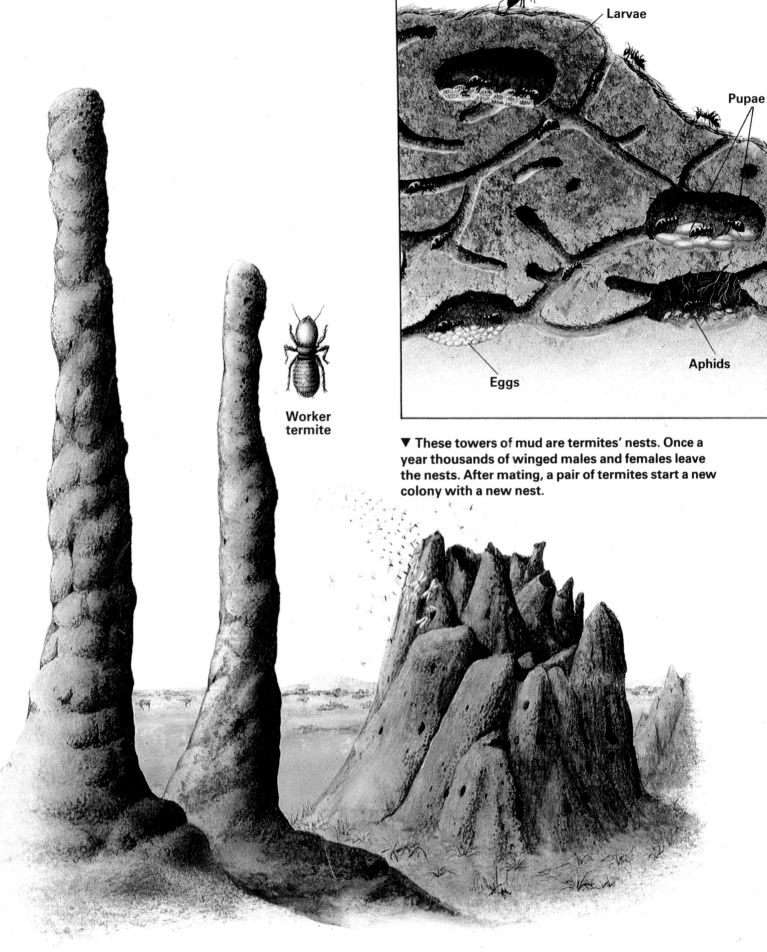

Worker termite

Larvae
Pupae
Aphids
Eggs

▼ These towers of mud are termites' nests. Once a year thousands of winged males and females leave the nests. After mating, a pair of termites start a new colony with a new nest.

◀ This wood ants' nest has been cut open to show different rooms. You can see where the eggs are stored, where larvae live, and where they stay when they pupate. In another room the ants keep aphids. The ants obtain honey dew from the aphids by stroking them with their antennae.

▶ A queen bee lays one egg in each of these wax cells made by the workers. After three days each egg hatches into a worm-like larva. The workers surround the queen to clean and feed her.

Thousands of people live and work in a city. Some do one kind of work, some another. The different jobs they all do help the city to run smoothly. It is like this with certain kinds of insects that live together in large numbers.

Honeybees are one example. As many as 50,000 honeybees may share one hive. Each bee performs a special task. Male bees are known as *drones*. Their work is simply mating with a queen bee on her wedding flight. A queen bee's job is laying eggs. Worker bees carry out all the other tasks.

Workers are females which do not lay eggs. They perform the housework in the hive and also act as nurses to the bee *larvae* or baby bees. They feed the larvae on *royal jelly*, a substance produced by their bodies. Later the larvae are given honey and pollen. Workers also guard the hive. They fly out of the nest to feed and collect nectar and pollen from flowers. The bees turn the nectar into honey and store it in wax *cells* or *combs*.

Like honeybees, ants live in groups known as *communities*. As with the honeybees, the males do no work, the queens lay eggs and the workers fetch food and guard the nest. Some kinds of ant have workers with large heads, armed with powerful jaws. These workers are the soldiers.

Most ants nest on the ground. The nests have different rooms. One is where the queen lays eggs. Another is where the eggs hatch into grubs. A further room is for the grubs when they turn into white *pupae*. From these hatch the males and future queens as well as the workers. Extra rooms are for storing food.

Termites are insects that look like ants. Some kinds build nests of soil as tall as a house. Inside there are many tunnels where the termites live. Each colony has a king and a queen. There are thousands of small white workers which dig tunnels and find food and water. They also feed the soldiers. Soldier termites have huge heads and jaws to defend the colony from enemies. Workers and soldiers are usually blind.

# Finding a Mate

When a male and female of the same kind meet and mate, the female produces young. Different animals find mates in different ways. Some female moths give off a scent that males can smell from far away. Male grasshoppers chirp to help females find them. Red deer stags bellow to invite female deer to come to them. The sounds also warn other males to keep away. If another stag does approach, the two animals fight with their antlers.

Once a male and female have seen each other, the male may use special ways of persuading her to come close. Each male fiddler crab has one huge brightly-colored claw. The male waves his claw to attract passing females to the little patch of muddy shore where he has his burrow.

Bright colors and big fins help some male fishes to find a mate. A male Siamese fighting fish has much larger fins than the female. Each male woos a female by spreading out his fins and waggling his body.

Many cock or male birds are more gaily colored than the hens or females. Some cock

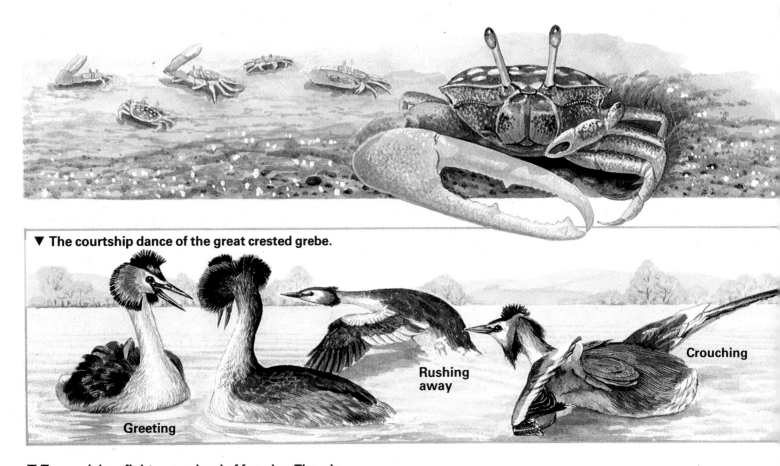

▼ The courtship dance of the great crested grebe.

Greeting  Rushing away  Crouching

▼ Two red deer fight over a herd of females. They do not usually hurt each other with their antlers. It is the one that tires first that loses the fight.

birds strut about or dance to show off their fine feathers. A peacock has beautiful long feathers growing from just above the tail. It spreads these out like a giant fan to win the attention of the peahen.

It is different with male and female great crested grebes, for both look much alike. Instead of just the male showing off his feathers, the two birds dance together in the water. They shake their heads and crouch and dive. Then the male stretches upwards. Sometimes both birds stretch and offer each other bits of water plant. Behavior like this is called *courtship*. Courtship brings the animals together and makes it possible for them to mate.

◀ **A male fiddler crab twitches its claw in a certain way to warn off other males. It will move the claw differently when it wants to attract a mate.**

Siamese fighting fish

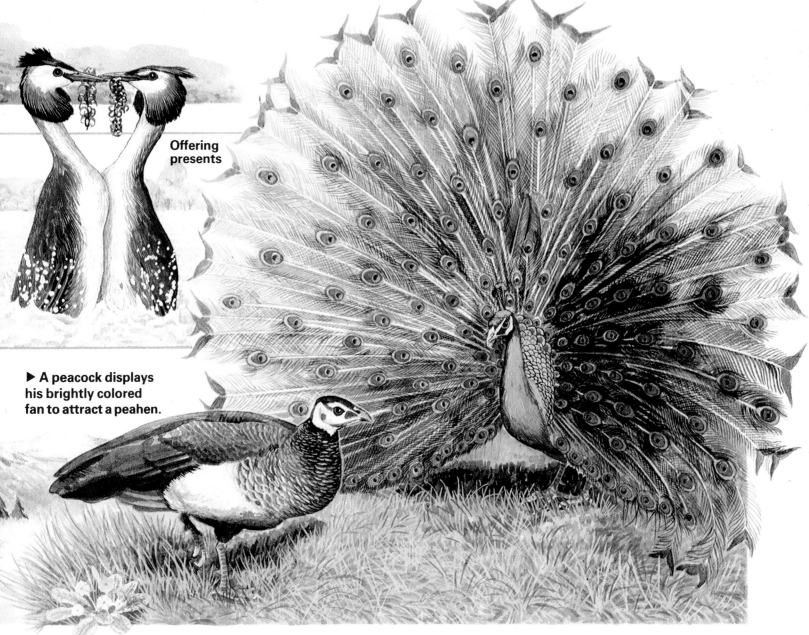

Offering presents

▶ **A peacock displays his brightly colored fan to attract a peahen.**

# Animal Homes

Big animals like whales and elephants have no homes. Neither do animals with hooves like horses and musk oxen. But tiny animals such as mice need somewhere safe to sleep and hide from their enemies. Creatures as small as wrens and as large as bears have nests or dens where they can bring up their young.

Some animals find homes ready-made. Hares rest in a *form*. This is just a dip in the ground. Bats sleep in caves or hollow trees, and a hollow branch is a home for a raccoon.

Other animals make homes where they can breed or shelter. Most have no hands like ours to help them, so they use whatever tools they have such as beaks, teeth and claws.

Harvest mice weave dry grasses into a hollow ball. Potter wasps mold mud into a pot-shaped nest. Weaver ants pull leaves together and glue them with a kind of silk. Foam-nesting tree frogs use their legs to beat jelly into a foam that keeps their eggs and tadpoles moist.

Rabbits and moles use their paws to dig long tunnels. Moles live underground all the time. Rabbits sleep, breed and hide in burrows. Only slender enemies like stoats can chase a rabbit down its hole. Even then, a rabbit may escape from its burrow by another entrance.

**Harvest mice**

**Potter wasp and nest**

**Weaver ants at work**
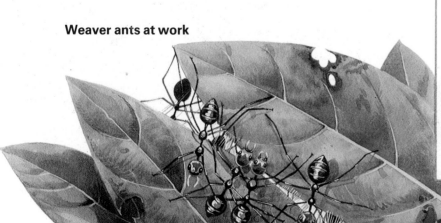

**Foam-nesting tree frog**

▲ Moles tunnel for worms. Their young are born in special nesting chambers.

▲ A rabbit burrow has several tunnels and rooms. One room usually serves as living quarters. The baby rabbits on the right are born in a separate burrow.

# Master Builders

Chimpanzees sleep in trees on platforms made of broken branches. But many birds and mammals make much more complicated nests.

A weaver bird starts by weaving an upright ring of grasses around growing plants. The bird sits inside the ring and goes on working until the ring becomes a pocket. Then the weaver closes the front of the ring. The finished nest looks rather like a ball. Its only entrance is a small hole underneath. This nest keeps out rain and most kinds of enemy. Hundreds of weaver birds' nests may hang from just one tree.

Oven birds get their name from nests that look like tiny, old-fashioned bakers' ovens. The rufous ovenbird builds on a fence post. It uses mud and adds tiny roots to make the mud stronger. The finished nest has a rounded roof and an entrance on one side. A twisting passage leads into the nest chamber. This little room is partly shut off to make it hard for enemies to find and reach the eggs.

Most mammals nest on or near the ground, but squirrels build high among the branches to avoid their enemies. Their nests, called *dreys*, are made of twigs and leaves packed tightly to shut out wind and rain.

Few mammals build as cleverly as beavers. They chop down small trees with their sharp teeth and cut them into smaller pieces. Then they block a stream with the wood and small stones. A pool collects behind this dam and the beavers build their nest or *lodge* there. First they pile a heap of branches, mud and stones in the pool. Then they gnaw openings into the heap from underneath. Inside they make a chamber above the water. Here they raise their young and spend the winter warm, dry and safe from enemies.

An ovenbird in the entrance to its nest of mud.

▼ Follow the stages in which the weaver bird builds its nest.

▲ Red squirrels feed and play outside their home, a rainproof nest of twigs. Nearby, a hole in the tree trunk provides a long-eared owl with a roosting place.

▼ This is what a beaver's home would look like sliced through the middle. The lodge stands in a pool made by the beaver's dam. The chamber inside can only be reached from underwater.

▶ This raccoon has made its home inside a hollow log. Raccoons also climb trees to make their dens in holes above ground. Others sleep and bring up their babies in rock crevices or on rocky ledges.

# Looking after their Young

Baby animals start life in one of two main ways. Creatures such as frogs, crocodiles and birds hatch from eggs laid by their mother. Most mammal babies, like monkeys and kangaroos, live inside their mothers' bodies until they are born.

Both kinds of baby come from tiny specks of jelly. Bit by bit the specks grow larger and start to take shape. We call these unhatched and unborn creatures *embryos*.

Embryos need food to make them grow. An embryo chick feeds on yolk inside its shell. Most embryo mammals take food from their mothers' bodies.

A growing embryo also needs warmth. Frog embryos get enough warmth from the water around them and many reptile eggs take warmth from the soil. But embryo birds and mammals need extra heat. Birds warm the embryos by sitting on their eggs. A mammal embryo keeps warm because it grows inside its mother's body.

Some animals can care for themselves almost from the moment they are born or hatched. A baby guppy can swim about, feed, and dodge enemies as soon as it leaves its mother's body. A baby partridge or chicken can run about on the day it is born.

Other young animals start life helpless. They would quickly die without parents there to feed them, shelter them from sun and rain and guard them from enemies waiting to eat them.

Most baby birds that live in nests depend on their parents to bring them food. Baby mammals feed on milk sucked from their mothers' bodies.

After a few weeks, many baby birds can find their own food. But some young mammals need their parents' help for much longer. A newborn kangaroo is one of the most helpless of mammal babies. When it is born it is only about an inch long – no bigger than a bumblebee. It spends about 200 days in the safety of its mother's belly pouch, drinking milk and growing. Baby chimpanzees are far larger than kangaroos when they are born. But years pass before they are old enough to look after themselves and leave their mother.

▲ When a baby hippo learns to walk, it often tries to move away from its mother. But its mother keeps nudging it back and the baby hippo soon learns to stay near her where she can protect it.

▶ The male midwife toad carries the female's eggs wrapped around its back legs.

▲ The herring gull broods its eggs in a simple nest. The young chicks peck at the red spot on the parent's beak to ask for food.

▼ A Japanese macaque cares for her baby. Baby monkeys learn how to look after themselves by copying their mothers and other adults.

▼ A baby kangaroo finds food and safety in its mother's pouch.

# The Winter Sleep

In winter, cold and hunger make life difficult for many animals.

Cold-blooded creatures such as snakes and lizards cool down with their surroundings and grow sluggish. If their surroundings freeze, their bodies also freeze and they die.

Warm-blooded animals like bears and dormice have fur to keep them warm. But fur only helps to stop their body heat leaking out. They need food to heat their bodies in the first place. In winter, frost kills many plants and insects, and snow may cover the ground. So certain foods grow scarce and difficult to find. Mammals that cannot find food to eat would soon use up all their energy and either freeze or starve to death.

Some animals escape cold and hunger by migrating (see pages 28-29). Others *hibernate*. Hibernation is a kind of deep sleep that may last all through the winter. Hibernating animals can survive for months without eating because their bodies use up so little energy.

Before they hibernate, animals store food in their bodies. In summer and early fall, toads and hedgehogs eat so much that they grow fat.

▲ Five animals from several lands are shown here in the places where they sleep for the winter. A butterfly has found a frost-free hibernating chamber in a hollow tree. A frog hides at the bottom of a pond, well below the icy surface.

While they hibernate, they gradually use up this fat in keeping their bodies just alive.

As the fall days grow cooler, animals ready for their winter sleep find hiding places that will be safe from winter frost or ice. Snakes and lizards hide in crevices or holes. Hamsters burrow. Dormice creep inside nests of moss and leaves. Bears use caves or hollow logs. Fishes may burrow in the mud at the bottom of a pond or lake. Hibernating frogs and terrapins can also spend winter under water without drowning. Their bodies take what little oxygen they need from air dissolved in the water.

Hibernating mammals become very cold. In freezing weather, a hibernating dormouse can cool down from its usual 37 degrees Centigrade to an almost icy 2 degrees. Its heart and lungs work so slowly that it seems dead. Even dropping the creature on the ground will not wake it.

In cold northern lands, hibernating animals may spend three-fourths of the year asleep. But not all animals that sleep in the winter are really hibernating. Bears and badgers doze, but do not become cold as the dormouse does. Badgers may sleep for days — or even weeks in the coldest parts — but in most places they are active throughout the winter. Other flesh-eating animals, like stoats and wolves, take no extra sleep in winter. They stay awake and prowl hungrily in search of food.

▲ Adders, lizards and hamsters hibernate under stones or in burrows.

▼ A brown bear dozes on the floor of a cave. It wakes from time to time to feed. Above its head, bats hibernate with their wings held close to their bodies.

# Animal Journeys

Each year millions of animals make long journeys to places where they feed or breed. These journeys are called *migrations*.

Some animal migrations are short. When snow covers the high mountain slopes, wild sheep and goats climb down to lower, greener slopes. Other migrations are longer. When drought kills the grass they eat, the big African grazing antelopes called wildebeeste may trek 100 miles or more to fresh pastures. When winter comes to northern North America, herds of caribou plod 1000 miles south to feed and shelter in the forests.

Brown lemmings are small furry rodents which live in the far north. Unlike the other animals shown here, lemmings do not make return journeys. They *emigrate* only when their numbers grow so great that there is not enough food in the area where they live. Crowds of them sometimes try to cross stretches of sea that are too wide and they drown.

At sea, fur seals, whales and fishes all migrate. Every year some whales swim 10,000 miles to browse in the cold but food-rich waters of the Arctic or Antarctic. Later they will swim 10,000 miles back to warmer waters to give birth to their calves. Gray whales migrate further than any other mammal.

Eels migrate only twice in their lives. As babies, they drift and swim across the Atlantic Ocean to Europe and North America. As adults they swim back to the Sargasso Sea to mate and lay their eggs. Then they die.

Terns and some other birds migrate further than any animal that travels on the ground or in water. In the fall, Arctic terns fly from the Arctic to the Antarctic – a journey half way around the world. In spring they fly back again.

Other birds, such as swallows and white storks, spend the winter in warm places like Africa and fly north to breed in Europe every spring. The birds sometimes fly in groups, sometimes alone. Often they cross mountain ranges and oceans. But they always arrive at the same places where their ancestors have gone for thousands of years. No one knows exactly how these birds find their way year after year.

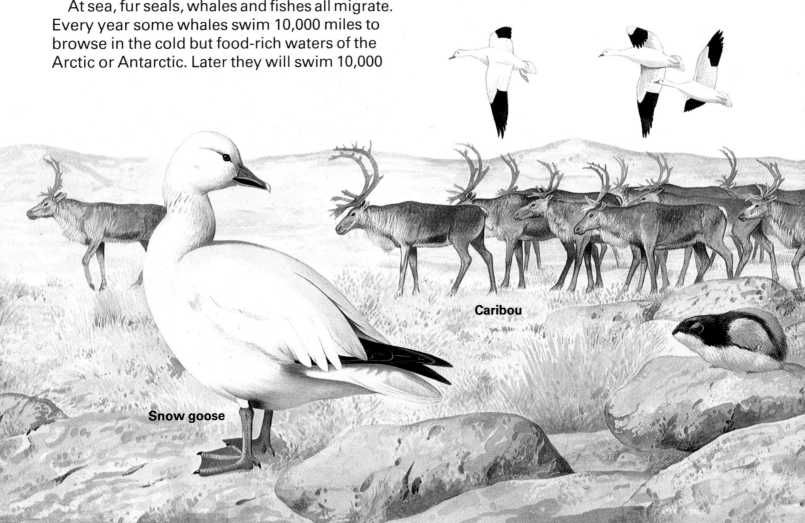
Snow goose
Caribou

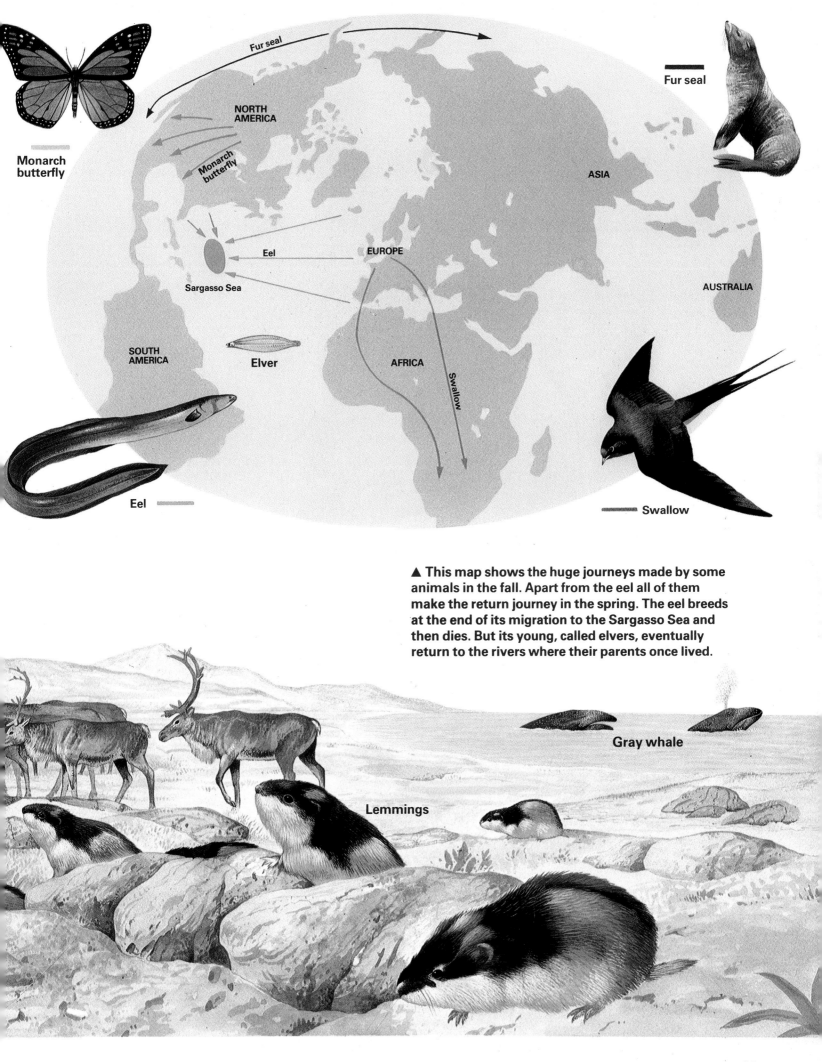

▲ This map shows the huge journeys made by some animals in the fall. Apart from the eel all of them make the return journey in the spring. The eel breeds at the end of its migration to the Sargasso Sea and then dies. But its young, called elvers, eventually return to the rivers where their parents once lived.

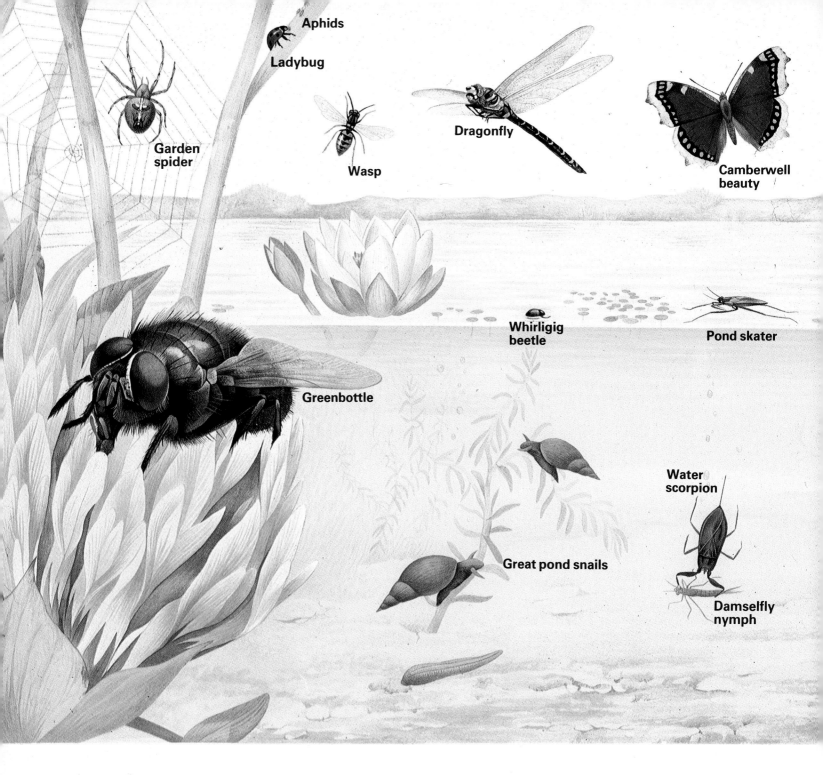

# Small Creatures

Air, soil and water teem with tiny creatures. Unlike fishes, birds and mammals, most of these little animals have no bony skeleton inside to help support their bodies. Some, like worms, slugs and snails, have no legs and lie flat and creep about slowly. Others, like insects and spiders, have jointed legs.

Insects have six legs; spiders have eight. Their legs and bodies are surrounded by a tough covering. This outside skeleton supports their bodies and protects them rather like a suit of armor. It also keeps their insides moist when the air outside is dry. So insects and spiders can live in drier places than worms and slugs which have to keep their soft skins damp.

As an insect grows, its outside skeleton becomes too small. So it splits and a new one takes its place. This happens several times during the insect's life.

Some insects also change shape as they grow. The young larvae may look quite unlike the full-grown insects. Caterpillars, for example, grow up into butterflies and moths.

Several of the creatures in the picture above depend on living plants for food. Aphids suck

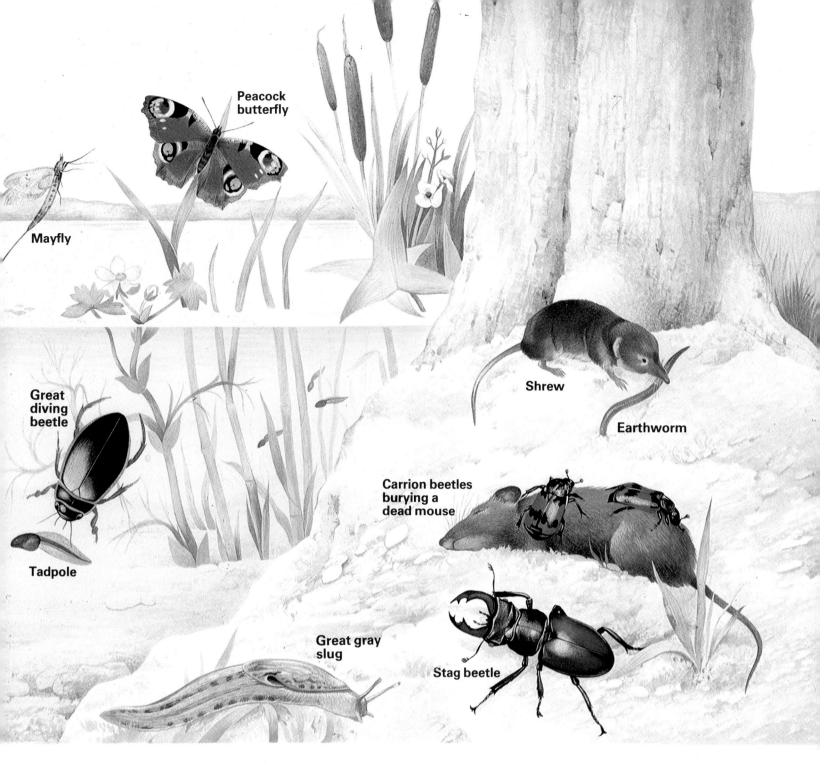

plant juices. Slugs eat green leaves and butterflies sip nectar from flowers.

Other small animals are hunters. Ladybugs eat aphids. Great diving beetles and water scorpions attack small fishes. Dragonflies, wasps and pond skaters hunt other insects.

One group of small creatures feeds on dead animals and plants. Carrion beetles eat dead birds and mice. Earthworms browse on rotting leaves in the soil.

In turn, all kinds of small creatures provide food for larger animals. Birds and shrews will gobble up worms, for example, while fishes snatch at flies.

▼ The trapdoor spider lives in a hole with a hinged lid. If small creatures come near its hole, the spider will leap out, pull them in and eat them.

# River Life

Just like a field or forest, a river is a larder stocked with food for animals that live there. Water snails, ducks and mammals such as water voles eat plants that grow under the water or sprout along the river bank. Worms, water fleas, freshwater shrimps and flies are gobbled up by fishes such as trout or bass.

Fishes in their turn are caught by larger hunters. Otters chase and catch them under water. Kingfishers dive down and pluck them from the surface. A long neck and a long beak help the heron catch fish from the bank. It also seizes frogs hunting insects in the grass.

Different creatures like different kinds of rivers. Some, including the trout and the dipper, like a clear, cool, fast-flowing mountain stream. The

▼ Here you can see some of the creatures that live in or beside the cool rivers of the north.

dipper is a small bird that actually walks and swims under water in search of small animals. Others prefer the muddy, sluggish salty waters where the river meets the sea. Many kinds of river animals live between the mountains and the sea, where the river winds its way across a plain. Here water plants can take root in the mud that settles thickly on the river bed and provides plenty of food for ducks and fishes.

The kinds of animals in the rivers also vary from country to country. In the warm rivers of the tropics you would find creatures like turtles and electric fishes. Because turtles are cold-blooded they need water warm enough to keep their bodies active. This is why many freshwater turtles live in hot countries. Tropical rivers are also the homes of the pompadour, the neon tetra and many other strange or brightly colored fishes that people catch and keep in aquariums.

▲ Arrow poison frogs live near rivers in South American rain forests.

▼ A snake bird spears a fish and Florida cooters swim in the warm waters of the southern states. Below them you can see some of the fishes that swim in the tropical Amazon River in South America.

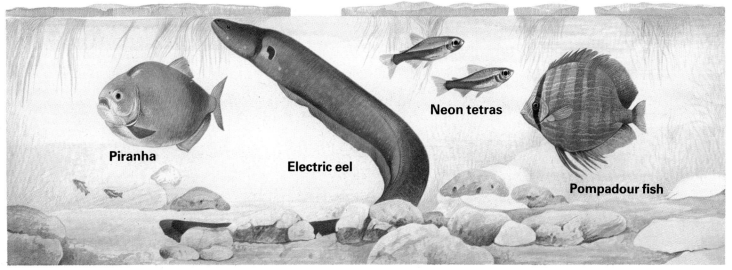

# Snakes and Lizards

Snakes, lizards, crocodiles and turtles are reptiles. Reptiles are cold-blooded animals and are active only when the Sun's heat warms their bodies. They have a scaly, waterproof skin. Most lay eggs with a waterproof shell. So reptiles can live and breed in lands too dry for moist-skinned amphibians like frogs which have to lay their thin-skinned eggs in water.

There are over 3000 species of lizard in the world. Many are very small. But the Komodo dragon can be over ten feet long and can kill and eat a pig.

Most lizards catch much smaller prey than this. Chameleons shoot out a long, sticky tongue to trap flies or other insects on the tip. Chameleons climb trees and bushes. But many lizards live on the ground. Some skinks burrow. Their smooth, shiny skins help them to push through sand.

When a lizard is attacked it usually runs away. But a frilled lizard tries to scare off enemies by stiffly stretching out its wide skin collar. If an enemy grabs a lizard by the tail, the tail may snap off and wriggle about. This gives the lizard time to escape. Later it will grow another shorter tail.

▲ Marine iguanas are the only lizards that swim in the sea. They live in the Galapagos Islands in the Pacific Ocean.

▲ The anaconda from South America is one of the world's largest snakes. The longest anaconda ever measured was over 25 feet.

Komodo dragon (Indonesia)

Gila monster (North and South America)

Gaboon viper (Africa)

▲ An Indian cobra's bite can kill a human in 15 minutes by stopping the heart and lungs from working.

Most lizards have four legs. Some have two. A few have none at all. Snakes have no legs. They move by wriggling along and pressing back on little bumps on the surface of the ground.

Unlike lizards, snakes can unhinge their jaws to swallow prey much larger than their own heads. All snakes are flesh-eaters. They eat animals such as rats, lizards, rabbits and birds. Constricting snakes like boas and pythons wrap themselves around their prey and squeeze until the victim suffocates. Poisonous snakes such as coral snakes and cobras have sharp hollow teeth, called *fangs*. When they bite, deadly poison squirts out through the fangs into their victim.

There are snakes that burrow, snakes that climb, even snakes that live entirely in the sea. There are at least 2300 kinds of snake.

Most snakes and lizards lay eggs covered in tough skin. Many bury their eggs in sand or soil where the Sun's heat hatches them. Grass snakes lay eggs in compost heaps so that they are warmed by the rotting vegetation. Reptiles' eggs take weeks or months to hatch. But common lizards and such snakes as vipers give birth to living young. A baby viper can give a poisonous bite as soon as it is born.

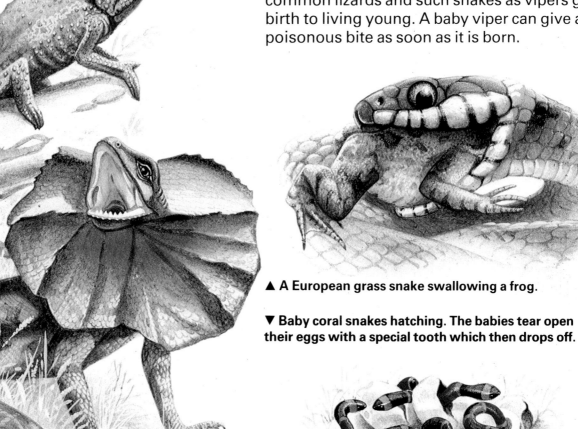

Chameleon (Africa)

Frilled lizard (Australia)

▲ A European grass snake swallowing a frog.

▼ Baby coral snakes hatching. The babies tear open their eggs with a special tooth which then drops off.

# Animals of the Dark

Many animals are only active in the dark. Most spend the day hiding from the Sun or from their enemies. Others live in caves or in the soil.

When night falls in a European wood, moist-skinned animals like frogs and newts come out to feed. For now the ground is cool and damp enough to stop their skins drying up. Mice, rabbits and other timid plant-eaters creep out of their holes at dusk. Insect-eating animals like hedgehogs crunch up beetles and snails. Bats wheel in the air to capture moths. Foxes prowl after rabbits and owls plunge down on mice and voles. An owl's soft feathers let it swoop silently through the air and seize its victim before it realizes there is any danger.

Different kinds of animals that feed by night live in different parts of the world. All of them can find their way about in light so dim that we could only grope our way around.

▶ The atlas moth is one of the largest moths in the world. Like the tarsier and the slow loris, it lives in South East Asia.

Slow loris

Atlas moth

Tarsier

Owls, tarsiers and slow lorises have huge eyes that can use the tiniest glimmer of light from the stars. Hedgehogs feel for food with their sensitive snouts. Mice and rabbits sniff the air to find their way to food. Bats can make 5000 high-pitched squeaks a second and find moths by the echoes bounced back from their tiny bodies. A pit viper can track down mice or birds by the heat that leaks out from their warm bodies.

As dawn arrives, night creatures hide. But earthworms, moles and mole-rats can feed day or night hidden in the darkness under the surface of the soil.

Some creatures live deep inside caves where it is always dark. Many of them are white in color. They either have tiny eyes or none at all and are quite blind. They find their food by smell or touch. Some cave salamanders and centipedes look like other salamanders and centipedes when they are born but their color fades as they live in their cave homes. You can see some of these animals on the page opposite.

▶ These animals feed by night in European woods.

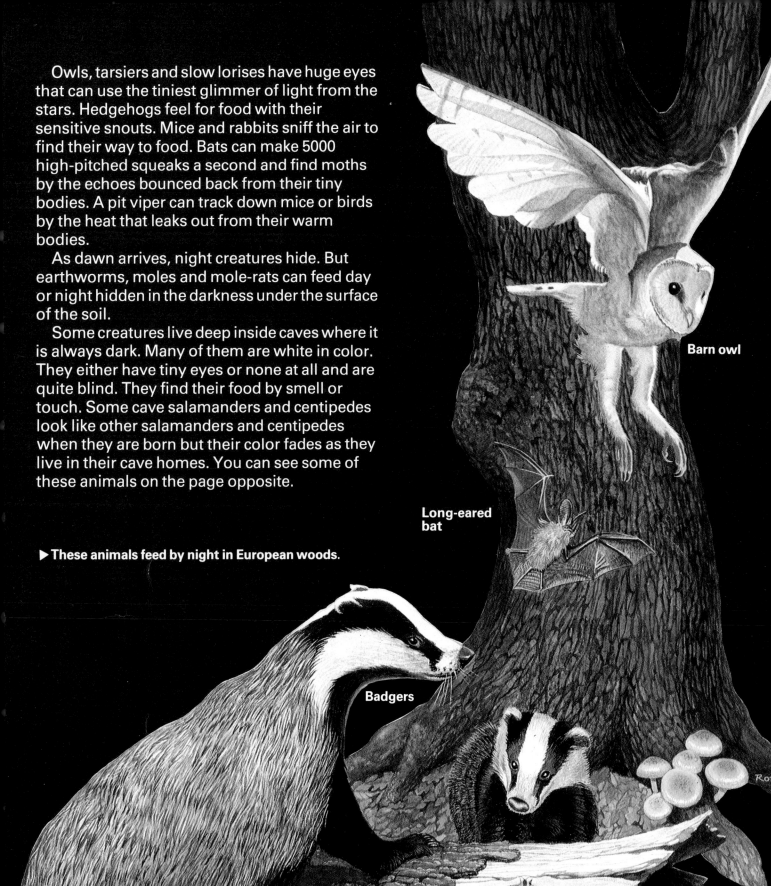

# Animals of Shallow Seas

Billions of small creatures live in the shallow water where the sea meets the land. In storms they are battered by waves and when the tide goes out many are left high and dry. Seashore creatures must be tough to survive these dangers. Most of them live on one particular type of shore.

Some animals live on rocky coasts. At low tide the rocks trap pools of water. Each rock pool may hold prawns, small crabs and tiny fishes. They can find food here and hide in crevices.

The rocks above the pool are the home of limpets, mussels, topshells, barnacles and sea anemones. They cling on tightly so that the waves cannot shift them. Waterproof shells help most rock dwellers to keep their bodies moist when the tide falls. Some sea anemones cover their bodies in waterproof slime. Little sea snails, called periwinkles, have a kind of lung which allows them to breathe out of water for months.

Most of the rock dwellers eat seaweed growing on the rocks or trap small particles of food brought in with the rising tide.

Sandy shores are quite different. They have no pools for animals to swim in nor rocks for them to cling to. Animals that live on sandy shores must burrow to escape storm waves and to avoid drying up when the tide goes out.

At high tide, mollusks like cockles, razorshells and tellins climb just under the surface of the sand. They send up feeding tubes that suck in water and particles of food.

Worms buried in sandy tubes push out tentacles that grope around for food scraps lying on the beach. Shrimps crawl out on to the sand to feed. Other small fishes swim in with the tide to hunt the shrimps, tube worms and mollusks. But lugworms are hard to catch. These buried worms do not come up to feed. They find food in the sandy and muddy soil they swallow.

▼ Only a few live animals, such as the shore crab, are found on sandy shores. The mermaid's purse is the washed-up egg case of a skate, a flat fish related to sharks. The empty shells once held soft-bodied mollusks, perhaps killed by fishes, birds or storms.

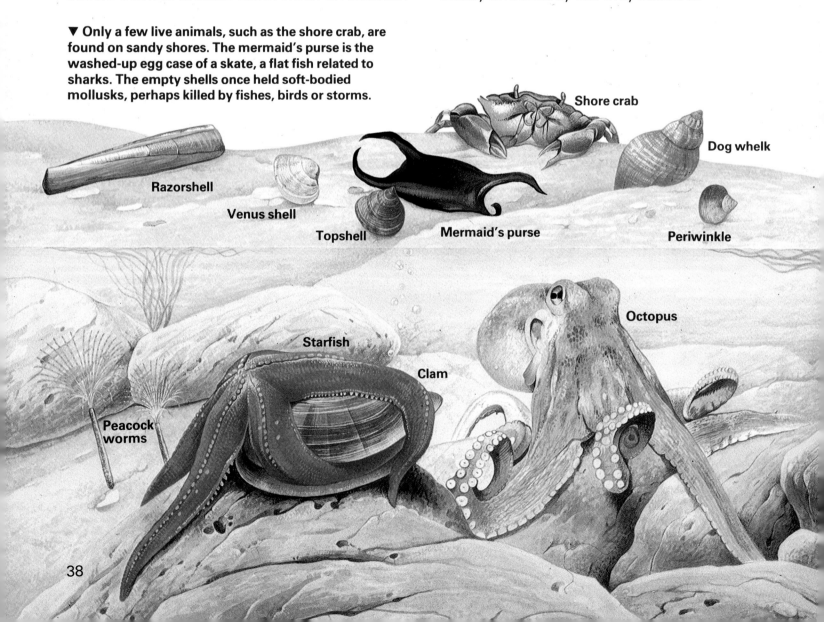

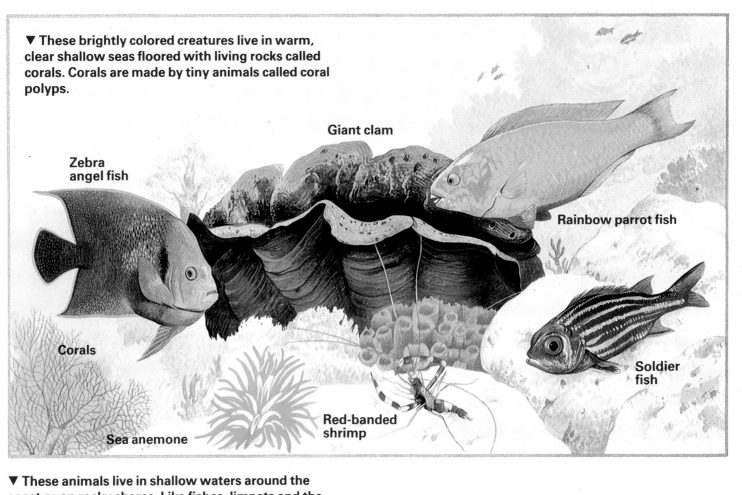

▼ These brightly colored creatures live in warm, clear shallow seas floored with living rocks called corals. Corals are made by tiny animals called coral polyps.

▼ These animals live in shallow waters around the coast or on rocky shores. Like fishes, limpets and the octopus, the starfish moves around to feed. The peacock worms, barnacles and mussels stay put and trap particles of food that float to them in the water.

# Ocean Animals

Ocean animals are well built for life in water. Most fishes have streamlined bodies shaped like torpedoes. They swim by waggling their tails, and brake and steer with their fins.

Fishes take in water through their mouths and let it out through slits in the sides of their heads. There are fleshy flaps inside these slits called *gills*. As water passes through a fish's head, the gills take in some of the oxygen dissolved in the water. This is how a fish breathes.

Most fishes contain a bag called a *swim bladder* which holds gas. As the bag fills with gas the fish floats. As the bag begins to empty the fish sinks. In this way the swim bladder helps a fish to swim at whatever level it wants in the water.

Many fishes have a bony skeleton. But sharks and rays have a skeleton of *cartilage* which is softer than bone. Sharks also lack swim bladders. They have to keep swimming to stop themselves sinking to the ocean floor.

**Giant squid**

**Leatherback turtle**

Like the fishes, the big sea mammals – whales, walruses and seals – are splendidly designed for swimming. But in other ways they are quite different. Firstly sea mammals are warm-blooded and have a thick layer of blubber which stops their body heat escaping into the cold water. Secondly, they must swim up to the surface of the water to breathe. But they can hold their breath for many minutes while they dive. Walruses and seals haul themselves ashore to have their babies but a mother whale gives birth at sea.

Fishes and sea mammals share the oceans with many other creatures. Some are large. The giant squid is the world's largest animal without a backbone. The leatherback turtle is a reptile weighing more than half a ton.

Small fishes as well as the largest whales depend for food on *plankton* – the mass of tiny, drifting plants and animals living near the surface of the oceans. In their turn, small fishes are food for larger kinds of fishes. Dead animals also provide food. They drift down and become meals for strange, little deep-sea fishes.

▲ The walrus mainly feeds on clams which it grubs out of the mud with its tusks.

▼ All the animals pictured here live in the oceans. Fishes the size of mackerel and herring are food for bigger, fiercer fishes like the great white shark and for sea mammals like the dolphins. The blue whale is the largest animal alive, yet it feeds on tiny shrimp-like animals called krill.

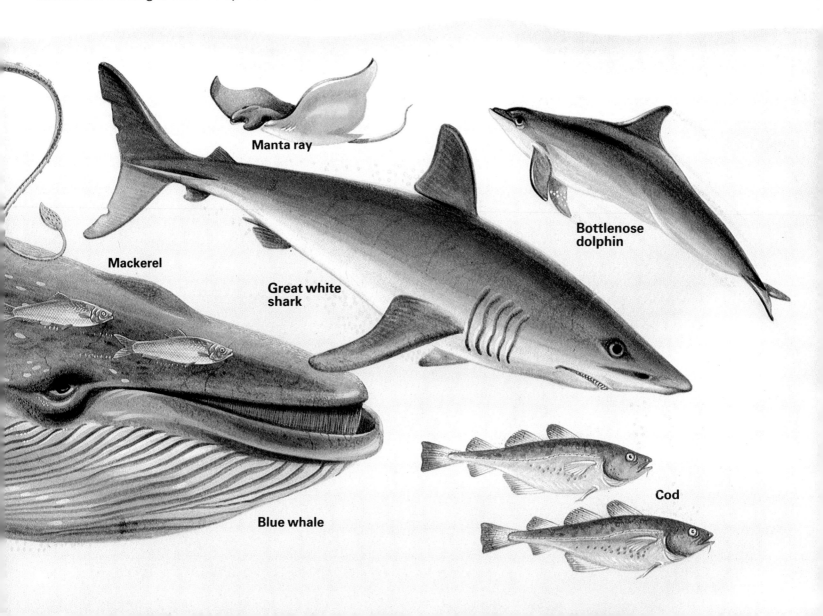

# All Kinds of Birds

▲ A spoonbill feeds in water. Behind, two crowned cranes perform a courtship dance. Like swans and some geese, crowned cranes mate for life.

▲ These seven woodland and forest birds come from many countries. Blackbirds, blue jays and crossbills live in cool northern lands. Birds of paradise, toucans and hummingbirds are at home in the hot rainy forests of the tropics. Pheasants come from southern Asia.

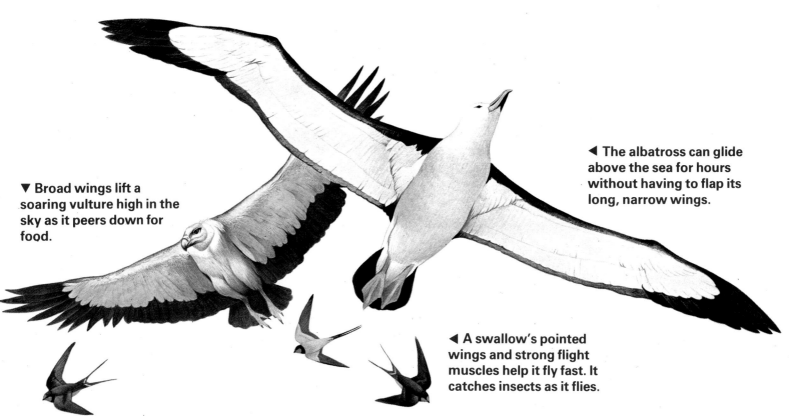

▼ Broad wings lift a soaring vulture high in the sky as it peers down for food.

◀ The albatross can glide above the sea for hours without having to flap its long, narrow wings.

◀ A swallow's pointed wings and strong flight muscles help it fly fast. It catches insects as it flies.

**Flightless Birds**

◀ The ostrich is the largest bird. It lives on the African plains.

▶ The kiwi lives in New Zealand. It hunts by night.

◀ An emperor penguin keeps its chick warm in the cold Antarctica.

There are about 9000 kinds of bird in the world. Because most birds can fly, they have been able to explore and settle in many parts of the world. Flying also helps birds to escape their enemies.

Most birds are built for flying. Their bones, feathers and beaks are light. Big strong muscles work their wings. Each beakful of air they breathe goes through their lungs twice to give them extra energy to flap their wings. Their stiff feathers are for flying and keeping out rain. Other soft feathers, called *down*, near a bird's skin keep it warm.

There are all sizes and shapes of birds, from hummingbirds the size of bumblebees to ostriches larger than people. Some birds like eagles and vultures have long broad wings for soaring on warm air currents. Others like gulls have long narrow wings for gliding. Ostriches and some other birds cannot fly at all.

Birds have many different kinds of beak too. Toucans have huge beaks that can reach fruit on slim twigs, while hummingbirds suck nectar from flowers with beaks like straws.

Each type of bird is at home in a special kind of place. Its shape, beak and feet allow it to find food and survive in its surroundings. Parrots eat fruit in tropical forests, while penguins fish in chilly ocean waters.

Most birds lay eggs and raise their young in nests which they have built themselves. But cuckoos lay their eggs in other birds' nests. These foster parents raise the cuckoos' chicks.

# Elephants

Elephants are the biggest living land animals. There are two kinds of elephant – Indian and African. Indian elephants are smaller than African elephants and have smaller ears. They live in southern and South East Asia. Some African elephants can grow up to 15 feet tall and weigh over six tons – more than the weight of 80 people.

Elephants have legs as thick as tree trunks to support their bulky bodies. They have thick skin but their hair is thin and short. Almost bare skin helps elephants to keep cool in the hot countries where they live and their big ears also give off heat as radiators do.

Elephants feed on grasses, leaves, small branches and fruit. Their trunks are their noses. But elephants use them like a hand to pull leaves and branches from trees and stuff them into their mouths. In one day an elephant can eat about 900 pounds of leaves and branches. They also use their trunks to squirt water or dust over themselves. They enjoy wallowing in water as you can see by looking at the African elephants in the picture below.

Here, too, you can see a baby elephant in a herd that is made up largely of its mother and its aunts. The mother keeps her baby near her as she walks to protect it from animals that might try to kill it. If an enemy approaches, she raises her trunk and screams a warning.

Chewing tough food wears down an elephant's teeth. New teeth grow as the old ones drop out. An elephant loses its last teeth when it is about 70. When it cannot chew, it dies of hunger.

Elephants have few enemies since most creatures cannot pierce their thick hides. A charging elephant like the one below on the right would scare even a lion for the elephant could trample the lion or stab it with its tusks.

People are the elephants' worst enemies. Every year hunters kill many of them with guns or poisoned arrows and saw off their ivory tusks to sell. Today elephants are becoming rarer.

▼ The elephant's nearest living relative is the rabbit-sized hyrax. Hyraxes have hoof-shaped nails and climb rocks or trees.

▼ Indian elephants are caught and tamed for work in southern Asia. Here they move logs with their tusks and trunks at the order of their riders, called mahouts.

# The Big Cats

Cats are fierce flesh-eating animals of the group called *carnivores*. The soft pads on their paws let them creep up quietly on prey. Long strong legs help them spring suddenly on their victims. They kill quickly with a bite from knife-sharp teeth, or a blow from a clawed paw. Domestic cats hunt rats and mice. But lions and tigers can easily kill a zebra or a cow.

Lions live on the grassy plains of Africa. Their tawny color makes them difficult to see in the tall, dry grass. Only the males have a dark shaggy mane. Lions often live in groups called *prides*. A pride of lions kills big grazing animals, including zebras and wildebeeste.

Leopards live in Africa and Asia. They often hide up trees where their spotted coats blend with patches of sunlight and shadow. When an antelope or other animal walks under the tree, the leopard leaps down to attack. A leopard is very strong for its size and can haul a dead antelope high up a tree out of reach of other carnivores. Snow leopards are paler than ordinary leopards. Their color helps to camouflage them in snow-covered mountains. Some leopards are all black and are known as black panthers.

Tigers are the largest cats of all. Their striped coats match the pattern of light and shade in the tall grasses where they hunt. All wild tigers live in Asia.

Most cats creep slowly up on their prey and pounce. Only the cheetah can run fast enough to chase its victim some distance and then kill it. Cheetahs can run at speeds of more than 50 miles an hour. They once lived on plains all over Africa and southern Asia. But today most of them live in eastern Africa.

Jaguars are big cats living in South America. Most of them look rather like the leopard but some are all black. The only other American big cat is the puma which looks rather like a lioness.

▶ A puma ready to spring on to its prey.

◀ A lion and a lioness doze out of the heat of the African sun.

▶ A tiger lies in long grass to escape the hottest part of the day.

▶ A leopard stretches before it starts to hunt.

▶ A cheetah carefully surveys the plains for suitable prey.

▶ A black jaguar in a South American river. Like tigers, jaguars are strong swimmers.

# Bears, Pandas and Koalas

Bears are big, strong mammals with a coat of shaggy fur. They look cuddly, but all of them are dangerous. Bears have large heads with powerful jaws and their broad flat feet are tipped with heavy claws. Bears walk on all fours, but they can stand on their hind legs to peer over rocks and bushes.

Like cats, bears belong to the flesh-eating group of mammals called carnivores. But most bears are not agile hunters like the cats. Instead of killing big animals, many bears eat a mixture of plant and animal food such as fishes, ants, eggs, honey, nuts, berries and roots. Some bears are completely vegetarian.

The largest bears are some of the brown bears that live in the northern forests of Europe and North America. A large North American brown bear can measure nearly 10 feet high. Bears of this size are the largest flesh-eaters living on land.

▼ The giant panda feeds mainly on bamboo shoots.

▼ Brown bears fish for salmon in a forest river.

◀ A polar bear waiting by a seal's breathing hole.

▼ A koala with her baby on her back.

Some polar bears are almost as huge as the largest brown bears. They have creamy coats which blend with the snow and ice of the far north. Polar bears are expert hunters. They can run fast enough to catch a reindeer and swim far out to sea to search for food. Polar bears often hunt seals by waiting near a hole in the ice. When a seal comes up to breathe, the bear swipes the seal with a paw and hauls it on to the ice to eat. Polar bears also feed on fishes and dead whales.

All other kinds of bears are smaller than the brown and polar bears. Black bears live in North America and spectacled bears in South America. Asia has Himalayan bears, sloth bears and sun bears. Sun bears are the smallest bears; fifteen of them weigh the same as one large brown bear.

The giant panda and the koala are bear-like mammals but they are not actually bears at all. Giant pandas belong to the raccoon family. They live on the slopes of mountains in China.

The koala is a mammal with a pouch like the kangaroo. Koalas live in Australia and they eat only certain kinds of eucalyptus leaves. They spend most of their time in eucalyptus trees eating the leaves and dozing among the branches.

▶ The spectacled bear lives in South America.

# Apes and Monkeys

Like dogs, cats and horses, apes and monkeys are mammals. But they are far more intelligent than most other mammals. Scientists put monkeys, apes and people in a group of mammals called *primates*. Primate means first and, in many ways, primates are ahead of all the other animals.

All apes and most monkeys live in the forests of hot lands. They feed mostly on fruit, leaves and insects and are well suited for finding food and safety among the trees. Their hands have fingers and thumbs that can grasp branches and sticks. Chimpanzees use sticks to hook leafy branches down to them or to knock fruit down to the ground.

The eyes of monkeys and apes look to the front and can focus on one object. This helps monkeys to judge distances when jumping from one branch to another. Most monkeys also have long tails to help them balance, steer and brake as they swing through the trees.

One group of monkeys lives in South and Central America. They are called the New World monkeys. Spider monkeys, woolly monkeys and howler monkeys are all New World monkeys. They can grip branches with their tails as well as their hands. South America is also the home of the tiny marmosets and tamarins.

Old World monkeys live in Africa and Asia. The colobus monkey can make enormous leaps from tree to tree as it seeks leaves to eat. Baboons are Old World monkeys that feed mainly on the ground. Old World monkeys tend to have narrower noses than New World monkeys and none of them can grip branches with their tails.

Unlike monkeys, apes have no tail and their arms are longer than their legs. Gibbons use their long arms to swing very fast from tree to tree. They are more nimble than the great apes – the chimpanzee, the orang utan and the gorilla. Great apes are the largest primates.

Lemurs belong to a different group of primates from the apes and the monkeys and are less intelligent. They have wet noses like dogs and a very keen sense of smell.

**Gorillas (Africa)**

# Animal Oddities

There are animals of all shapes and sizes in the world and they all have their own special way of living. Some animals seem particularly strange and a few of these are pictured here.

Sloths are slow-moving mammals that spend their entire lives hanging upside down from branches. They are almost helpless if they fall on to the ground.

Narwhals are small whales. The male narwhal has a front tooth that grows into a long tusk.

The aardvark looks like a pig with a long, tube-shaped snout and donkey's ears. It claws open termite nests and it licks up the insects inside with its long sticky tongue.

The duck-billed platypus lives in Australian rivers. It is a mammal but seems to be part bird too. A mother platypus lays eggs like a bird but feeds her babies on milk like a mammal.

Someone once said that a seahorse has a horse's head, a monkey's tail and a kangaroo's pouch. In fact, seahorses are tiny fishes that swim upright. Each female lays her eggs in a pouch in the male's body.

Tuataras and ajolotes are unusual reptiles. Tuataras look like lizards but they are the only living beakheads. Their ancestors lived millions of years ago in the Age of Reptiles. Tuataras can walk about in air cold enough to put most other reptiles to sleep. And they can go for up to an hour without breathing.

Apart from its two tiny limbs, an ajolote looks like a worm. Its eyes are hidden under the skin which is folded into rings. Like worms, ajolotes burrow underground. They are not worms but reptiles of a group called worm lizards.

Flamingos are among the strangest of all birds. They have longer necks and legs for their size than any other bird. Their beaks are strangely bent and they hold their heads upside down to feed in shallow muddy waters.

Many creatures without backbones seem very strange to us. But few are stranger than some sea cucumbers. If an enemy attacks they squirt out their insides. These entangle the enemy and the sea cucumbers escape. Later the lost parts regrow.

However odd they may seem to us, the strangest animals in the world have developed just that way to help them survive.

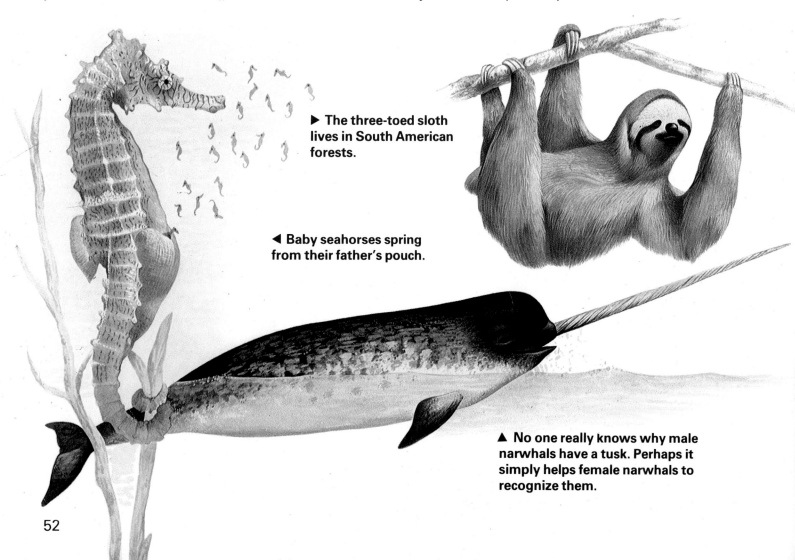

▶ The three-toed sloth lives in South American forests.

◀ Baby seahorses spring from their father's pouch.

▲ No one really knows why male narwhals have a tusk. Perhaps it simply helps female narwhals to recognize them.

▶ A sea cucumber.

▲ The worm-like ajolote is a reptile that burrows underground.

▼ The tuatara lives only on some islands off the coast of New Zealand.

▼ Africa's aardvark escapes its enemies by burrowing. One was seen to dig faster than a team of men with spades.

▲ Flamingos live in colonies in warm, shallow sea lagoons and salt lakes.

▼ The platypus has fur like other mammals but webbed feet and a beak like a duck.

53

# Animals of Long Ago

Millions of different kinds of animal have lived since time began. But many died out long ago. We know about them because parts of their bodies have survived in rocks as fossils. Many fossils are bits of skeleton turned to stone.

The creatures pictured on these pages lived scores of millions of years ago. They are all reptiles which developed from fishes and amphibians over millions of years. *Dimetrodon* was one of the earliest reptiles. It lived more than 250 million years ago in what is now Texas. *Dimetrodon* was a flesh-eater. It had sharp teeth and a skin "sail" that worked like a radiator. When it stood sideways to the Sun, the sail helped it to warm up. When it faced the Sun, heat escaping from the sail cooled it.

The other animals pictured here lived in the Age of Dinosaurs. This lasted from about 225 million years ago to 65 million years ago.

*Stegosaurus* was a plant-eater. It had bony plates jutting from its back to protect it. Perhaps the plates served as radiators too.

*Triceratops*

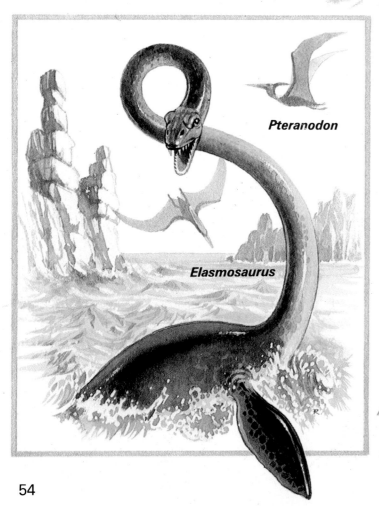

*Pteranodon*

*Elasmosaurus*

*Archaeopteryx*

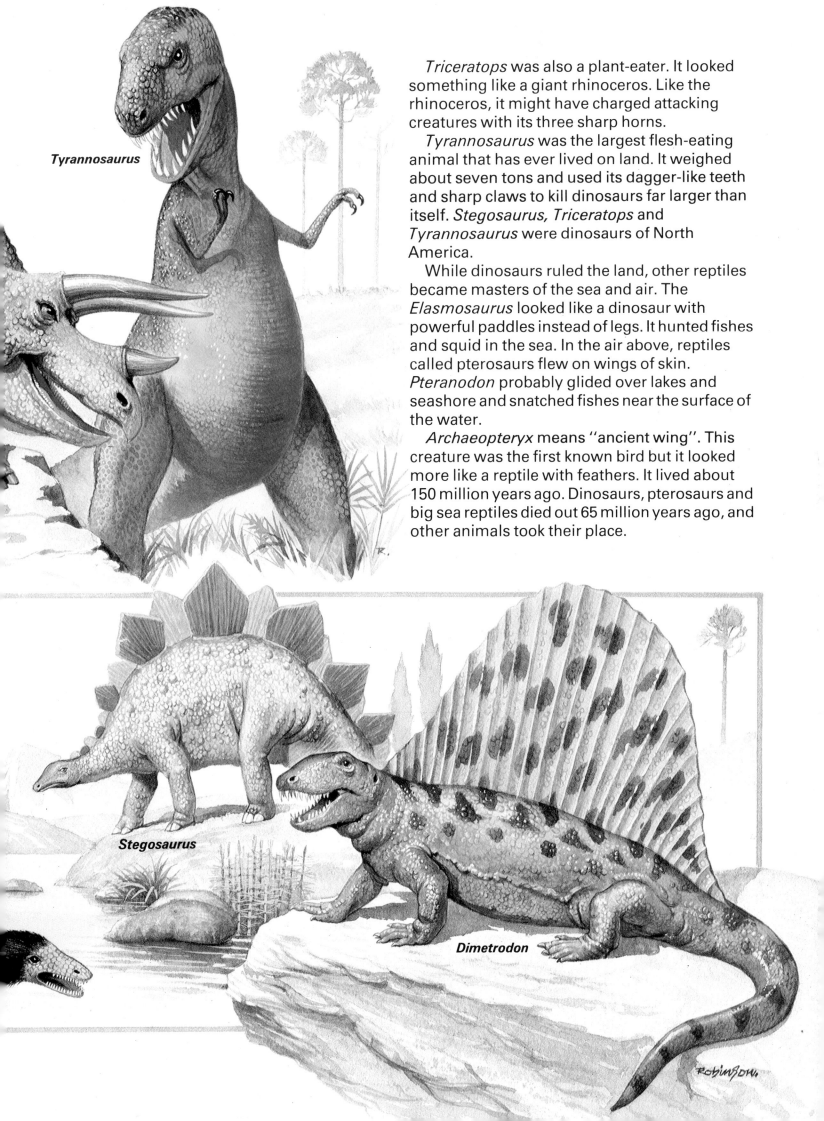

*Triceratops* was also a plant-eater. It looked something like a giant rhinoceros. Like the rhinoceros, it might have charged attacking creatures with its three sharp horns.

*Tyrannosaurus* was the largest flesh-eating animal that has ever lived on land. It weighed about seven tons and used its dagger-like teeth and sharp claws to kill dinosaurs far larger than itself. *Stegosaurus, Triceratops* and *Tyrannosaurus* were dinosaurs of North America.

While dinosaurs ruled the land, other reptiles became masters of the sea and air. The *Elasmosaurus* looked like a dinosaur with powerful paddles instead of legs. It hunted fishes and squid in the sea. In the air above, reptiles called pterosaurs flew on wings of skin. *Pteranodon* probably glided over lakes and seashore and snatched fishes near the surface of the water.

*Archaeopteryx* means "ancient wing". This creature was the first known bird but it looked more like a reptile with feathers. It lived about 150 million years ago. Dinosaurs, pterosaurs and big sea reptiles died out 65 million years ago, and other animals took their place.

# The Animal Kingdom

More than a million kinds of animal live on Earth today and they are all relatives. They all come from a few very simple ancestors which lived in water hundreds of millions of years ago. These were probably tiny specks of jelly with a single body cell, like the protozoa. From the first protozoon came simple animals like sponges, made up of many cells. Simple animals gave rise to more complicated ones with different kinds of cells that worked in special ways, as muscles and nerves for example. Such animals included jellyfish, worms, molluscs, arthropods and starfish. Fishes were the first backboned creatures, or *vertebrates*. And over the years, amphibians, reptiles, birds and mammals – the other backboned animals – have developed.

To make it easier to study animals, scientists have arranged all animals into groups. On these pages you will find some of the most important groups of animals.

## INVERTEBRATES

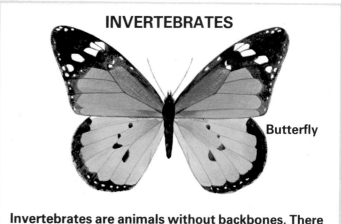
Butterfly

**Invertebrates are animals without backbones. There are hundreds of groups of invertebrates. Animals belonging to some of these groups are shown on this page.**

### Many-legged Creatures

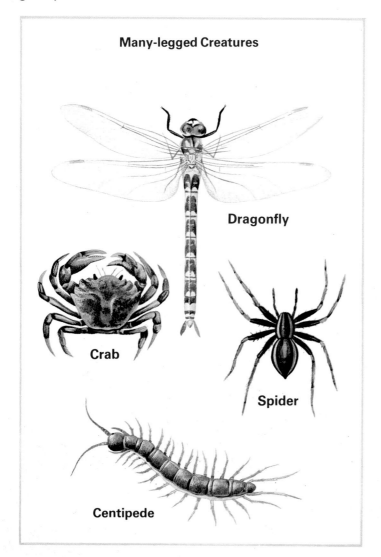

Dragonfly
Crab
Spider
Centipede

### Very Simple Creatures

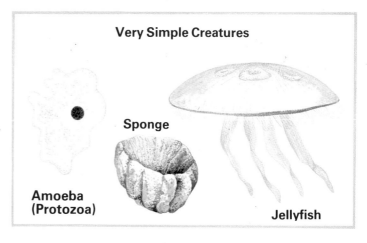

Amoeba (Protozoa)
Sponge
Jellyfish

### All Kinds of Worms

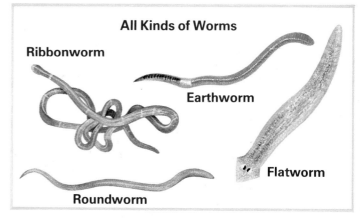

Ribbonworm
Earthworm
Flatworm
Roundworm

### Other Invertebrates

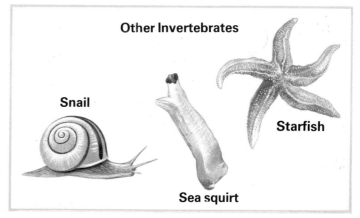

Snail
Sea squirt
Starfish

# VERTEBRATES

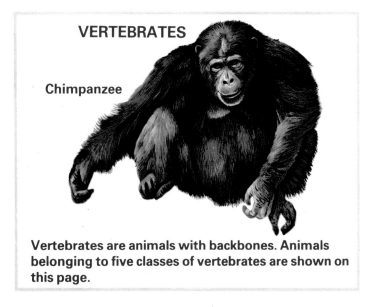

Chimpanzee

Vertebrates are animals with backbones. Animals belonging to five classes of vertebrates are shown on this page.

BIRDS are warm-blooded and lay eggs. They all have feathers and wings though not all birds fly.

Toucan

Ostrich

FISHES live in water. They have scaly skins and fins. Most kinds lay eggs.

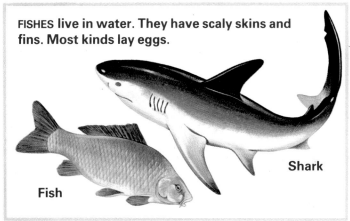

Shark

Fish

MAMMALS are warm-blooded and have hair on their bodies. Baby mammals drink milk from their mothers. Mammals are the most intelligent animals.

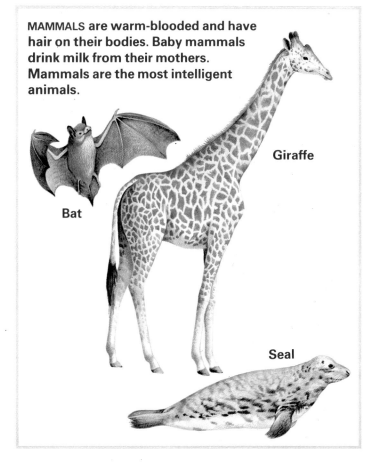

Giraffe

Bat

Seal

AMPHIBIANS are cold-blooded and lay eggs. They can live both in water and on land.

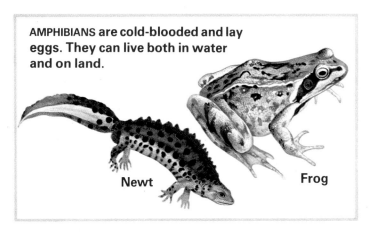

Newt

Frog

REPTILES have scaly skins and are cold-blooded. Most kinds lay eggs and live on land.

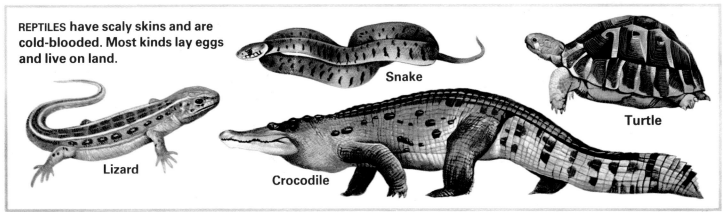

Snake

Turtle

Lizard

Crocodile

# Quiz

Test your memory on these fifty questions. You will find that some are easy but others are quite tricky. Page numbers are given before the questions so that you can go back to the right pages for help. The answers are on page 59.

**Pages 8 and 9**
1  Name the fastest animal in the world.
2  Name the fastest animal in water.

**Pages 10 and 11**
3  Which small fish has teeth as sharp as knives?
4  Why is an Arctic fox difficult to see in winter?
5  How does an armadillo defend itself?

**Pages 12 and 13**
6  Where are a cricket's 'ears'?
7  Which fox has the largest ears?
8  Name two animals that use echoes to find their way about.

**Pages 14 and 15**
9  What are parasites?
10  How do wild hunting dogs kill animals much larger than themselves?

**Pages 16 and 17**
11  How many honeybees may share one hive?
12  Which creatures look like blind white ants?
13  What does a queen ant do?

**Pages 20 and 21**
14  Why do some tree frogs beat jelly into foam?
15  What is a hare's home called?

**Pages 22 and 23**
16  What is a beaver's home called?
17  Which animals live in dreys?

**Pages 24 and 25**
18  What does an embryo chick live on?
19  How big is a newborn kangaroo?

**Pages 28 and 29**
20  How far may some whales migrate altogether in one year?
21  Where do swallows spend the winter?

**Pages 30 and 31**
22  Which beetles will eat dead mice and birds?
23  How many legs do insects have?

**Pages 32 and 33**
24 What do otters hunt?
25 What is a pompadour?
26 Where does the arrow poison frog live?

**Pages 34 and 35**
27 How many kinds of lizard are there?
28 Why can a snake swallow prey much larger than its own head?

**Pages 36 and 37**
29 How many sounds can a bat make in one second?
30 Name three animals that feed at night.
31 Where does the tarsier live?

**Pages 40 and 41**
32 Why would a shark sink if it stopped swimming?
33 Which is the largest animal without a backbone?
34 Name two sea mammals.

**Pages 42 and 43**
35 Name the largest bird.
36 How many kinds of bird are there?

**Pages 44 and 45**
37 How many people may weigh as much as one large African elephant?
38 What happens when an elephant loses its last teeth?
39 What is an elephant's trunk?

**Pages 46 and 47**
40 Where do leopards live?
41 What are all-black leopards called?

**Pages 48 and 49**
42 Which are the largest bears?
43 Where do pandas live in the wild?

**Pages 50 and 51**
44 What can a spider monkey do with its tail?
45 Do apes have tails?
46 Name the group of mammals that includes apes, monkeys and people.

**Pages 52 and 53**
47 Where does a female seahorse lay her eggs?
48 Which sea creature squirts out its insides to tangle up its enemies?

**Pages 54 and 55**
49 What helped *Dimetrodon* to warm up and cool down?
50 Which was the largest-ever flesh-eating land animal?

# Index

**A**
Aardvark  52, 53
Africa  14, 15, 28, 34, 35, 43, 44, 45, 46, 47, 50, 51, 53
Ajolote  52, 53
Albatross  43
America, North  28, 29, 33, 35, 46, 48, 49, 55
America, South  11, 33, 34, 46, 47, 49, 50, 51, 52
Amoeba  56
Amphibian  57
Anaconda  34
Antelope  9, 10, 11, 14, 28
Ant lion  10
Ant  16-17, 20
Antarctic  28
Ape  50-51
*Archaeopteryx*  54, 55
Arctic  28
Arctic fox  10, 11
Armadillo  10
Asia  35, 36, 42, 44, 45, 46, 48, 49, 50, 51, 52
Atlas moth  36
Australia  35, 49

**B**
Baboon  10, 50, 51
Badger  26, 27, 37
Barnacle  38, 39
Bat  12, 13, 20, 27, 36, 37, 57
Beak  43
Bear  26, 27, 48-49
Beaver  22, 23
Bee  17
Beetle  30, 31
Bird  8, 9, 11, 12, 18, 19, 22, 24, 28, 31, 42-43
Bird of paradise  42
Black bear  49
Blackbird  42
Bluejay  42
Blue whale  40-41
Brown bear  48-49
Buffalo  14, 15
Butterfly  26, 30, 31, 56

**C**
Caribou  28-29
Carnivores  44-45, 48, 54-55
Cat, wild  9, 10, 11, 44-45
Caterpillar  9, 10, 30
Cave animals  36, 37
Cave centipede  36, 37
Centipede  36, 56
Chameleon  34, 35
Cheetah  9, 10, 11, 44, 45
Chimpanzee  22, 24, 50, 51, 57
China  49
Clam  8, 38, 39
Clownfish  14, 15
Cobra  35
Colobus monkey  51

Coral  14, 39
Coral snake  35
Courtship  18-19
Crab  12, 18, 38, 56
Cricket  12
Crocodile  14, 15, 57
Crossbill  42
Crowned crane  42
Cuckoo  43

**D**
Deer  10, 18
*Dimetrodon*  54, 55
Dinosaur  54-55
Dipper  32, 33
Dog  8, 12, 14
Dolphin  41
Dormouse  26, 27
Dragonfly  9, 30, 31, 56

**E**
Eagle  10, 11, 43
Ears  12, 13
Earthworm  12, 31, 37, 56
Eel  28, 29, 33
Egg  16, 17, 20, 22, 24, 25, 34, 35, 43
*Elasmosaurus*  54, 55
Elephant  8, 20, 46-47
Embryo  24
Emperor penguin  43
Europe  28, 35, 38, 39
Eye  12

**F**
Feathers  43
Fennec fox  13
Fiddler crab  18
Fish  8, 9, 11, 13, 14, 15, 18, 19, 24, 27, 28, 29, 32-33, 36, 38-39, 40-41, 52, 57
Flamingo  52, 53
Flatworm  12, 56
Fly  9, 12, 13, 30, 31
Foam-nesting tree frog  20
Fossils  54
Fox  10, 11, 13, 36
Frilled lizard  34, 35
Frog  20, 24, 25, 26, 27, 32, 33, 36, 57
Fur seal  28, 29

**G**
Giant panda  48, 49
Giant squid  40, 41
Giant tortoise  8
Gibbon  50, 51
Gila monster  34
Gills  40
Giraffe  8, 57
Golden eagle  11
Gorilla  50
Goshawk  12
Grasshopper  18
Grass snake  35
Great crested grebe  18, 19
Grey whale  28, 29
Griffon vulture  43
Guppy  24

**H**
Hamster  27
Hare  8, 9, 20
Harvest mouse  20
Hedgehog  26, 36, 37
Heron  32
Herring gull  25
Hibernation  26-27
Hippopotamus  24
Horse  8, 9, 20
Hummingbird  42, 43
Hunters  10, 11, 14, 31, 32, 33, 44-45, 49
Hyrax  46

**I**
Iguana  34
Impala  11
Insect  9, 12, 13, 16-17, 18, 20, 26, 30-31
Invertebrate  56

**J**
Jaguar  44, 45
Japanese macaque  25
Jellyfish  56

**K**
Kangaroo  8, 24, 25
Kingfisher  32
Kiwi  43
Koala  48-49
Komodo dragon  34
Krill  41
Kudu  15

**L**
Ladybird  30, 31
Leatherback turtle  40, 41
Leech  14
Lemming  28, 29
Lemur  50, 51
Leopard  44, 45
Limpet  38, 39
Lion  44
Lizard  34-35, 57
Looper caterpillar  9

**M**
Mammal  24, 41, 50, 51, 57
Manta ray  41
Midwife toad  25
Migration  26, 28-29
Mole  20, 21, 37
Monarch butterfly  29
Monkey  25, 50-51
Moon moth  13
Moth  13, 18, 30, 36, 37
Mouse  20, 26, 27, 36, 37
Muscles  8, 43
Mussel  38

**N**
Narwhal  52
Nerves  13
Nests  16, 17, 20, 21, 22-23
Newt  57

New World monkeys  50
New Zealand  53

**O**
Octopus  38
Old World monkeys  50
Orang utan  50, 51
Ostrich  14, 43, 57
Otter  32
Oven bird  22
Owl  23, 36, 37
Oxpecker  14
Oyster  8

**P**
Panda  48-49
Parasite  14
Peacock  19
Penguin  43
Periwinkle  38
Pheasant  42
Piranha  11, 33
Plankton  41
Platypus  52, 53
Plover  14, 15
Polar bear  48, 49
Pond skater  30, 31
Porcupine  10
Porpoise  12
Potter wasp  20
Primate  50-51
Pronghorn antelope  9
Protozoa  56
*Pteranodon*  54, 55
Pterosaurs  55
Puma  44

**R**
Rabbit  20, 21, 36, 37
Racoon  20, 23, 49
Rattlesnake  13
Ray  40, 41
Razorshell  38
Red deer  18
Reptile  14, 15, 24, 34-35, 40, 41, 52, 53, 54-55, 57
Rhinoceros  10, 14
Ribbonworm  56
Roundworm  56

**S**
Sailfish  9
Salamander  36, 37
Sea anemone  14, 15, 38, 39
Sea cucumber  52, 53
Seahorse  53
Seal  28, 29, 41, 57
Sea lion  8
Seashore animals  38, 39
Sea squirt  56
Seaweed  38, 39
Shark  40, 41, 57
Shellfish  38-39
Shrew  31
Siamese fighting fish  18, 19
Skink  34
Sloth  52
Slow loris  36, 37
Slug  8, 30, 31
Snail  8, 32, 38, 56
Snakebird  33
Snake  13, 26, 27, 34-35, 57
Snow goose  28
Spectacled bear  49
Spider monkey  50, 51
Spiders  30, 31, 56
Spine-tailed swift  9
Spoonbill  42
Sponge  8, 56
Squid  8, 40, 41
Squirrel  12, 22, 23
Starfish  38, 56
*Stegosaurus*  54, 55
Stoat  14, 20, 27
Sun bear  49
Swallow  28, 29, 43
Swim bladder  40

**T**
Tarsier  36, 37
Termite  16, 17
Tern  14, 15, 28
Tiger  44
Toad  25, 26
Topshell  38
Tortoise  8, 10
Toucan  42, 43, 57
*Triceratops*  54, 55
Trout  32
Tuatara  52, 53
Turtle  33, 40, 41
*Tyrannosaurus*  55

**V**
Vertebrate  56
Viper  13, 34, 35, 37
Vulture  43

**W**
Walrus  41
Wasp  30, 31
Water scorpion  30, 31
Weaver ant  20
Weaver bird  22
Whale  20, 28, 29, 40-41, 52
Wildebeeste  14, 28
Wing  43
Wolf  10, 14, 27
Woodchuck  26
Worm  8, 12, 14, 30, 31, 37, 38, 56

**Z**
Zebra  8, 14

---

### ANSWERS TO QUIZ

1. Spine-tailed swift
2. Sail fish
3. Piranha
4. Because its white coat matches the snow
5. By curling into a ball
6. On its knees
7. Fennec fox
8. Porpoises and bats
9. Animals and plants that feed on others
10. By hunting in packs
11. 50,000
12. Termites
13. Lay eggs
14. To keep their eggs and tadpoles moist
15. A form
16. A lodge
17. Squirrels
18. Egg yolk
19. About two centimeters, the size of a bumblebee
20. 20,000 miles
21. Africa
22. Carrion beetles
23. Six
24. Fishes
25. A freshwater tropical fish
26. South America
27. Over 3000
28. Because it can unhinge its jaws
29. 5000
30. Badgers, owls, hedgehogs, bats, moths, mice, tarsiers, slow lorises
31. South East Asia
32. Because it has no swim bladder
33. A giant squid
34. Whales, dolphin, seal, walrus
35. Ostrich
36. About 9000
37. 80
38. It starves
39. Its nose
40. In Africa and Asia
41. Black panthers
42. Some of the brown bears
43. China
44. Grip branches
45. No
46. Primates
47. In a pouch in the belly of the male seahorse
48. A sea cucumber
49. A skin sail on its back
50. *Tyrannosaurus*.